YOUR KNOWLEDGE HAS VALUE

Bibliographic information published by the German National Library:

The German National Library lists this publication in the National Bibliography; detailed bibliographic data are available on the Internet at http://dnb.dnb.de .

Imprint:

Copyright © 2019 GRIN Verlag
Print and binding: Books on Demand GmbH, Norderstedt Germany
ISBN: 9783668995710

This book at GRIN:

https://www.grin.com/document/492456

Anonym

Aus der Reihe: e-fellows.net stipendiaten-wissen

e-fellows.net (Hrsg.)

Band 3152

Non-Covalent Catalysis and Hydrogen Bonding

GRIN Verlag

GRIN - Your knowledge has value

Since its foundation in 1998, GRIN has specialized in publishing academic texts by students, college teachers and other academics as e-book and printed book. The website www.grin.com is an ideal platform for presenting term papers, final papers, scientific essays, dissertations and specialist books.

Visit us on the internet:

http://www.grin.com/

http://www.facebook.com/grincom

http://www.twitter.com/grin_com

Experimental Organic Chemistry

Faculty of Mathematics and Natural Sciences

University of Cologne

Abbreviation

$CDCl_3$	deuterated chloroform
conc.	concentrated
d	doublet
DCM	dichloromethane
eq./ equiv	equivalent
et al.	and others
g	gram
h	hours
J	coupling constant
cat.	catalyst
LUMO	lowest unoccupied molecule orbital
M	molar mass
m	multiplet
MHz	megahertz
min	minute
ml	millilitre
mol	amount of substance
NMR	nuclear magnetic resonance spectroscopy
ppm	parts per million
rt	room temperature
s	singlet
T	temperature
t	triplet
THF	tetrahydrofuran
δ	chemical shift
λ	wavelength

Contents

1 Introduction

Nowadays it is common to use catalysis in organic synthesis. It can help in orienting the substrates, lowering barriers to reaction and accelerating the rates of reaction. In addition to metal-ligand systems and biocatalysts, there is another class of catalysts – the organocatalysts which are free of any metals; like many enzymes. The organocatalysts often consist of chiral compounds. The output materials are easy to find in the nature. How these catalysts accelerate the reaction rates is steady a central question in organic synthesis. It is important to distinguish the interactions with the organic substrates between covalent and non-covalent bonds. The activation of a carbonyl compound by conversion into an enamine or into an iminium ion belongs to the covalent catalysis while to increase the electrophilicity of a carbonyl group by formation of hydrogen bondings is a typical example for non-covalent organocatalysis. Thus, the acceleration and the control of the reaction rates depend on formation of hydrogen bonds for non-covalent organocatalysis.[1] It is possible to catalyse two hydrogen bonds which occurs in dual hydrogen bonding donors. This area and the asymmetric catalysis have received more attention in the last time.[2]

2 Knowledge

2.1 Hydrogen-Bond Catalysis

The idea of catalyzing by hydrogen bonds was created by imitating nature. Hydrogen bonds are needed in enzymes to maintain their structure and functionality. These biological molecules catalyze chemical reactions and even those that form carbon-carbon bonds. The binding of ligands to receptors is one of the illustrations of the noncovalent interaction. The hydrogen-bond is defined as an interaction between a proton donor and an acceptor molecule. Chemists developed chiral catalysts that use hydrogen bonds to achieve high enantioselectivity. In many asymmetric synthesis, chiral organic small-molecule catalysts are used which implicated hydrogen-bonding to an electrophile to activate it.[3] During the activation of the electrophile by the acidic catalyst, which forms hydrogen bonds to, for example, carbonyl compounds or imines, the LUMO energy levels of the C=O or C=N bonds are lowered, which may lead to a facilitated nucleophilic attack.[4] However, the catalysts differ in their types and in their application. Therefore, it is possible to classify the catalysts in different mechanistic classes. [3]

2.1.1 Dual Hydrogen-Bonding Donor Catalysis

In contrast to one hydrogen bonding, the interaction with two hydrogen bonds to a substrate is naturally stronger. Delivery of two bonds also proved to be a successful way to activate electrophiles in enzymes.[1] With forming two hydrogen bonds to a reactant also the geometry will be defined.[5] The electrophiles may consist of aldehydes, ketones, esters, imine derivatives, N-acyliminium ions and nitro compounds.[1] Chiral urea and thiourea derivates act as asymmetric catalysts.[2] Derived from them, new catalyst systems were developed, such as squaramides, P-triamides and cyclodiphosphazanes (figure 1).[5]

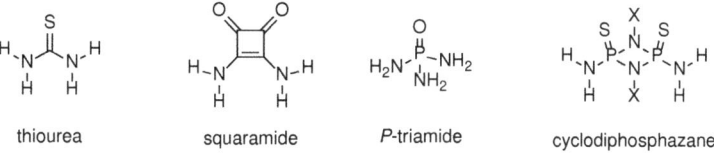

Figure 1. Structures of thiourea, squaramide, P-triamide and cyclodiphosphazane.[5]

2.1.2 Bifunctional Catalyst System

An interesting study revealed the simultaneous activation of electrophile and nucleophile in a reaction called bifunctional catalysis. Squaramides, for example, often have an additional chiral scaffold which carries a basic unit. This base can be protonate and build a new hydrogen bond (figure 2).[6] The bifunctional catalysts also include proline and cinchona alkaloid.[1]

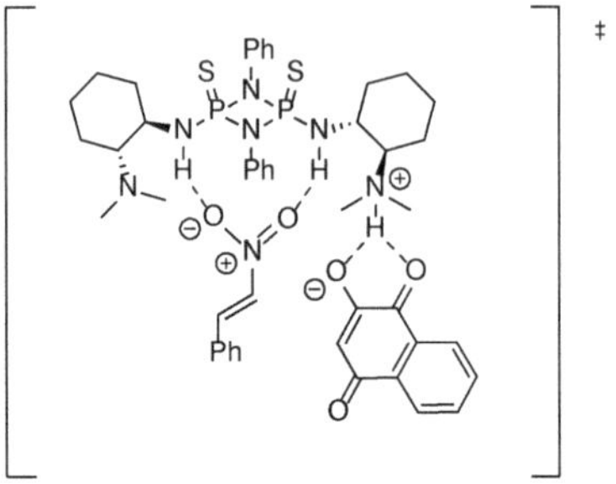

Figure 2. Transition state arising from the addition of 2-hydroxynaphthoquinone to β-nitro-styrene.

2.1.3 Single Hydrogen-Bond Donation

It is also possible to catalyze reactions with only one hydrogen bond, but this mechanism is much rarer than the double H-bond donation or bifunctional catalysis.[1] A simple chiral alcohol can use the hydrogen bond to catalyze important cycloaddition reactions of a diene with various aldehydes. TADDOL, a chiral diol, also forms hydrogen bonds with carbonyls, therefore it is often used in enantioselective reductions of ketones as in hetero-Diels-Alder reactions as a catalyst (figure 3).[7] In fact, there are structural similarities between TADDOL derivatives and phosphoric acids, which can also form a single hydrogen-bond for highly enantioselective catalysis.[1]

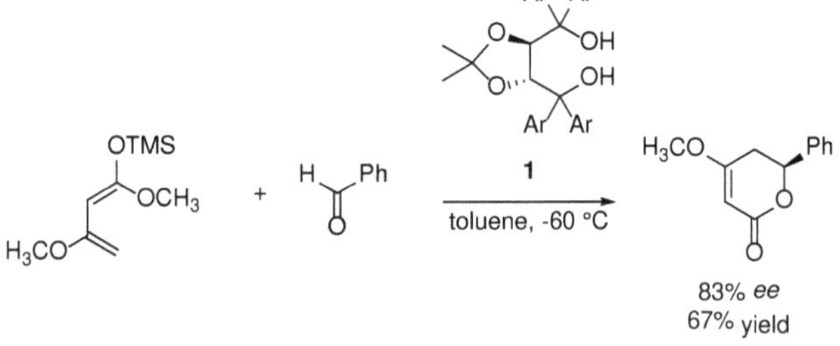

Figure 3. Asymmetric hetero-Diels-Alder reaction of diene with aldehyde catalysed by 10 mol% TADDOL derivate **1** (Ar = 1-naphthyl).[1]

2.2 Cyclodiphosphazane as Catalyst

As briefly shown above, cyclodiphosphazanes are four-membered rings having alternating phosphorus (V) and nitrogen centers. This compound makes it possible to promote hydrogen-bonding catalysis.[8] They are synthesized from dichloro-cyclodiphosph(III)azane and can then be equipped with ligand systems. The precursor is obtained by commercially available primary amines and PCl_3 (figure 4).[9]

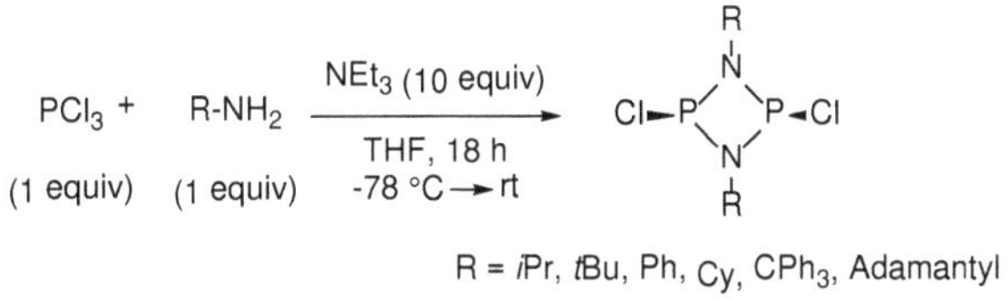

R = iPr, tBu, Ph, Cy, CPh₃, Adamantyl

Figure 4. General synthesis of dichlorocyclodiphosphazanes.[9]

As described, it is possible to form hydrogen-bonds by means of certain ligands. According to *Klare* et al. chiral primary amines were used as ligands. The product may specifically have different substituents or be symmetrically synthesized. In addition, the phosphorus atoms were oxidized by sulfur and cyclodiphosph(V)azanes were obtained (figure 5).[5]

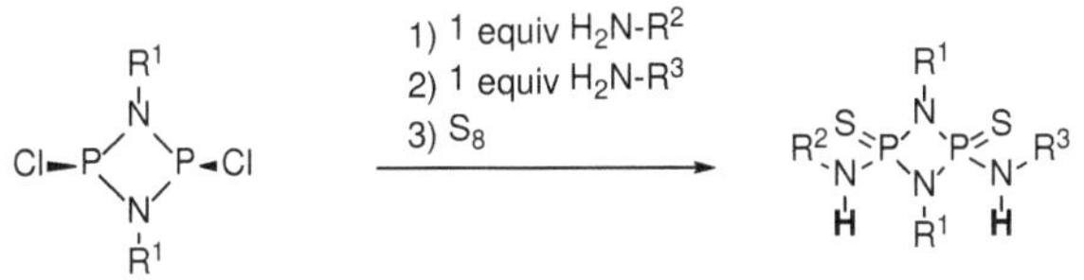

Figure 5. General synthesis of chiral cyclodiphosphazane catalysts.[5] Protons which are responsible for hydrogen-bonds are marked in bold.

They succeeded in producing different catalysts (figure 6). The catalysts carry chiral DMDACH or aniline as ligands.

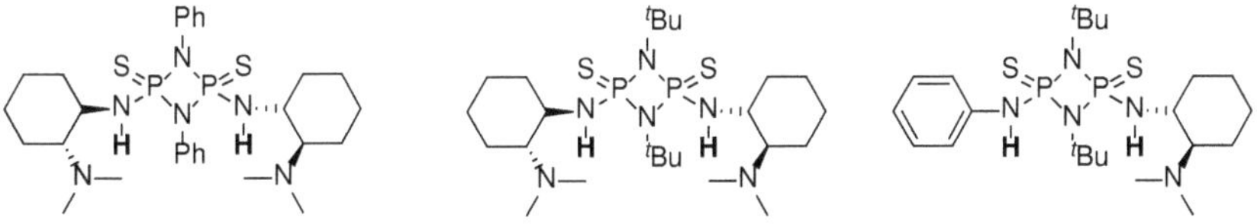

cis/ trans-**2a/2b**   all-cis-**3**   all-cis-**4**

Figure 6. Synthesized catalyst according *Klare* et al.[5] Protons which could be responsible for hydrogen-bonds are marked in bold.

4

Their potential to form hydrogen-bond was investigated in an asymmetric Michael addition. All synthesized catalysts probably act as hydrogen-bond donors. However, the cis-configured symmetric catalyst proved to be the most efficient (figure 7). It was assumed that it catalyzes the reaction in a bifunctional mechanism.

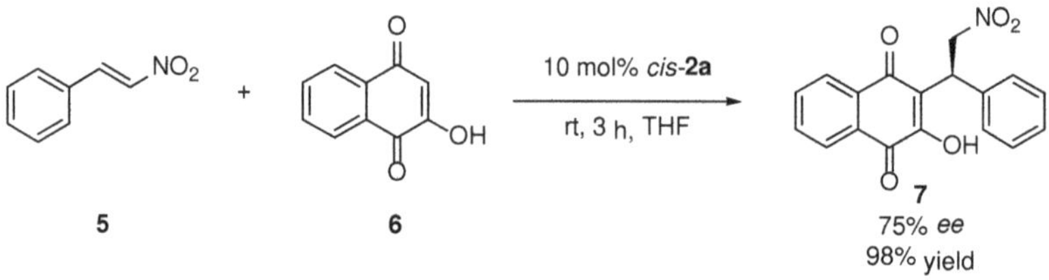

Figure 7. Asymmetric Michael addition reaction catalysed by *cis*-**2a** according *Klare* et al.[5]

At the amines of the catalyst, hydrogen-bonds are generated to the nitro group of the olefin. This increases the nucleophilic attack of the olefin. The tertiary amine is available as a base and is protonated by 2-hydroxy-1,4-naphthoquinone. Thus, the 1,4-dioxo-1,4-dihydronaphthalene-2-olate is formed, which acts as a nucleophile and can attack the electrophilic site of β-nitrostyrene.

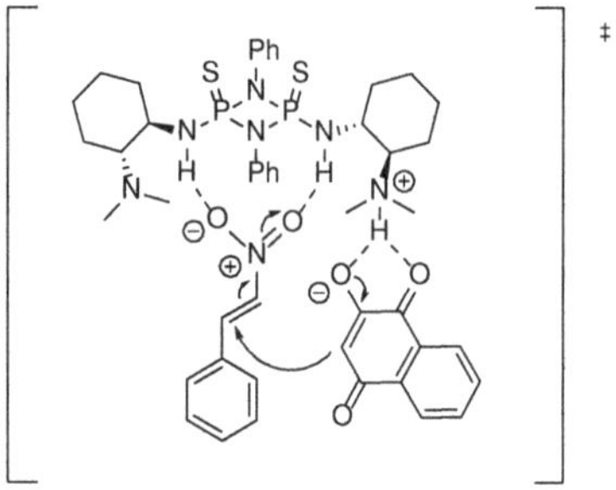

Figure 8. Transition state arising from the addition of 2-hydroxynaphthoquinone to β-nitro-styrene. [5]

Thus, in the transition state to both substrates hydrogen-bonds were formed over the catalyst (figure 8). Once through the diamide structure of the cyclodiphosphazane and on the other by the quaternary amine. [5]

3 Aim

Since the application in the hydrogen-bonding catalysis of the chiral cyclodiphosph(V)azane as a catalyst was successful[5], new catalysts with the same basic building block should be developed (figure 9). So only the ligands should be exchanged.

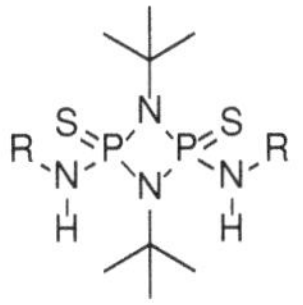

Figure 9. Structure of the basic building block of the developed catalyst.

In *Klare* et al. the Michael addition of 2-hydroxynaphthoquinone to β-nitro-styrene was investigated with a catalyst which undermine the bifunctional mechanism. In this work the same addition should be examine, thus the new catalyst has to meet the same requirements. Therefore, it was necessary to select ligands such that amino groups are available for hydrogen-bonding to the nitro group and a tertiary amine can act as a base. Since only one framework acts as a base, it is possible to synthesize an asymmetric catalyst. Thus, 3,5-bis(trifluoromethyl)aniline and the chiral amin-cinchonidine were used as ligands (figure 10). In addition, the phosphorus should be oxidized by both sulfur and oxygen. To optimize the catalysis, only the solvents should be changed.

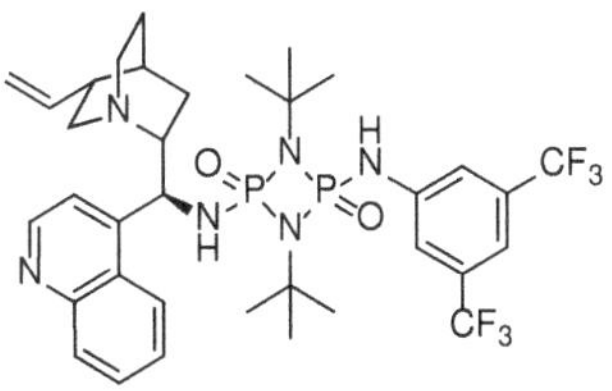

8 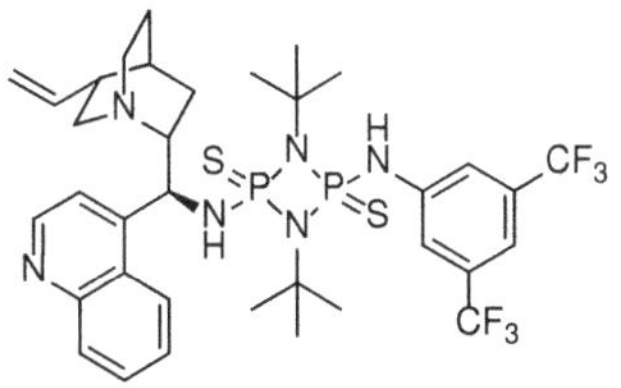   9

Figure 10. Target molecules **8** and **9** of this work.

6

4 Results and Discussion

4.1 Synthesis of Precursors

4.1.1 Synthesis of Chiral Amin Ligand

The synthesis of 9-amino-(9-deoxy)epi-cinchonidine **11** was carried out via a modified protocol of *Wan* et al. starting from the hydroxylated form of cinchonidine **10** (figure 11).[10] First, a Mitsunobu reaction with DPPA produced 9-azido(9-deoxy)epi-cinchonidine, which was then converted directly into the corresponding amine by a Staudinger reaction. The product was purified by column chromatography through dichloromethane, methanol and ammonia and 54 % of the desired chiral amine was obtained, which was characterized by ^{1}H-NMR spectra (see appendices). The reaction was carried out under dry conditions.

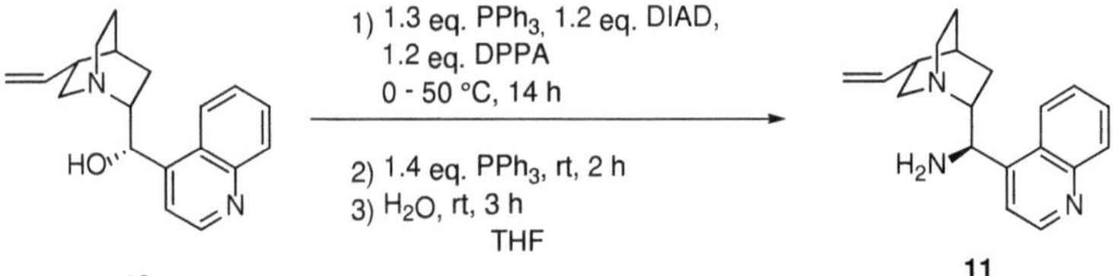

Figure 11. Synthesis pathway for the preparation of 9-amino-(9-deoxy)epi-cinchonidine **11**.[10]

4.1.2 Synthesis of Dichlorocyclodiphosph(III)azane

The synthesis of dichlorocyclodiphosph(III)azane **12** was carried out according to a protocol of *Bashall* et al. from phosphorus trichloride and tributylamine under dry conditions (figure 12).[11] The crude product was filtered off and freed from the solvent in an oil vacuum. The product was recovered via vacuum distillation. The residue was suspended in *n*-pentane and kept cool. Thus, 15 % of the desired product was obtained and characterized by ^{31}P-NMR spectra (see appendices).

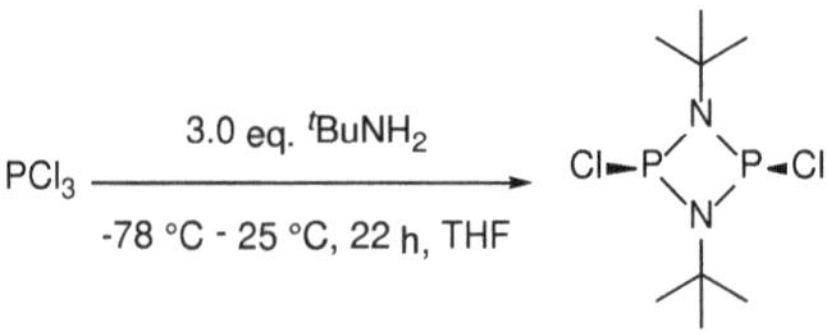

Figure 12. Synthesis pathway for the preparation of dichlorocyclodiphosph(III)azane **12**.[11]

4.2 Synthesis and Characterization of Catalysts

Braun also attempted to synthesize the asymmetric catalysts **8** and **9**, but to no avail, so that the regulation was changed.[12] He synthesized it analogous as in the synthesis of *Klare* et al. proceed.[5] For this purpose, the dichlorocyclo-diphosph(III)azane **12** was first reacted under dry conditions with one equivalent of 3,5-bis(trifluoro)aniline and cinchonidine in succession. The deprotonation of the amines was carried out via triethylamine. The difference with *Braun* was that aniline and triethylamine were not mixed together for adding rather added one after the other. As well during addition of the aniline, *Braun* cooled down the batch to a temperature of -78 °C and allowed the solution to stir for 24 h at room temperature, while in this work the batch was cooled to 0 °C upon addition of the aniline and was stirred for 24 h at 35 °C. After subsequent filtration, the resulting intermediate **13** was reacted with hydrogen peroxide or sulfur to oxidize the phosphorus (III) to phosphorus (V). The product was then purified by column chromatography with dichloromethane, methanol and concentrated ammonia. 24% of the oxygen oxidized **8** (from now o-cat) and 6% of the sulfur oxidized **9** (from now s-cat) product were obtained (figure 13). Both products could be successfully characterized (^{1}H-NMR, ^{31}P-NMR, ^{19}F-NMR, HRMS - see appendices).

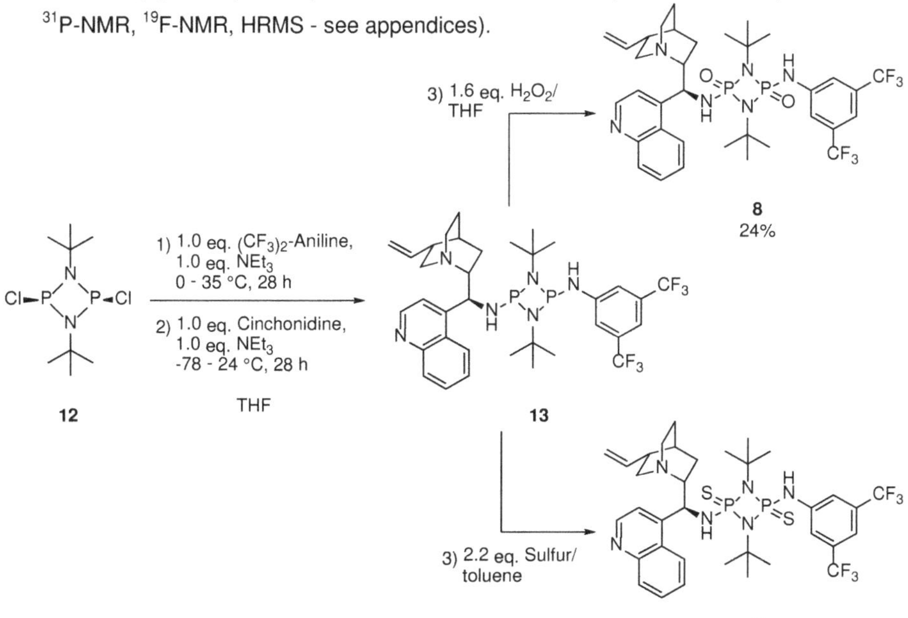

Figure 13. Synthesis pathway for the both catalysts **8** (o-cat) and **9** (s-cat).[11]

8

The low yields are not satisfactory and can still be optimized. Both products were purified by column chromatography, but the by-products have similar R_f values, therefore only a small amount could be cleanly isolated. The other fractions were mixtures.

Above all, the analysis of $^{31}P\{^1H\}$-NMR spectra was also used for turnover controls. Between the individual steps of the synthesis ^{31}P-NMR spectra were recorded to check whether the starting material has completely reacted with the individual ligands. For the dichlorocyclodiphosph(III)azane, the signal is mainly characterizing at 207 ppm. With complete disappearance of the signal, it can be assumed that a complete conversion happened. Figures 14 and 15 show two $^{31}P\{^1H\}$-NMR spectra. In this case, the dichlorocyclodiphosph(III)azane is to be seen in the upper and in the lower in each case the catalyst contained oxygen or sulfur.

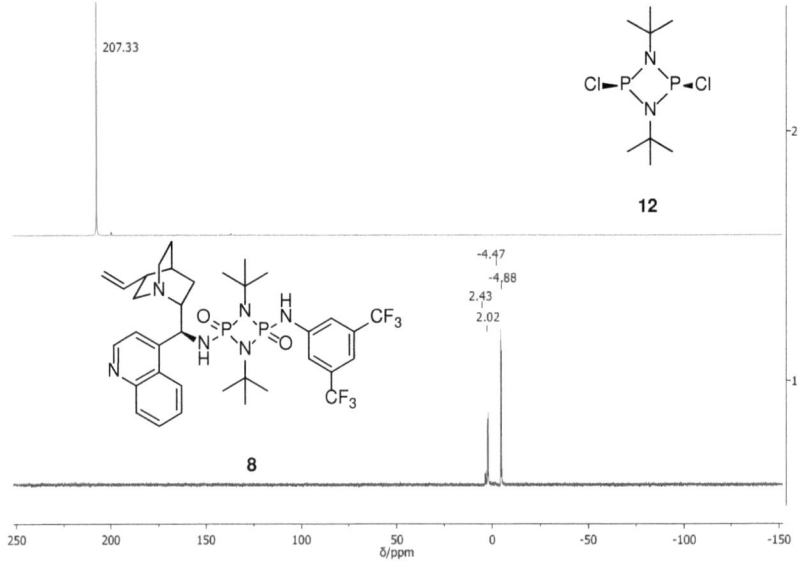

Figure 14. $^{31}P\{^1H\}$-NMR spectra (50 MHz) of dichlorocyclodiphosph(III)azane **12** compared to the catalyst with oxygen (o-cat) **8** in CDCl$_3$.

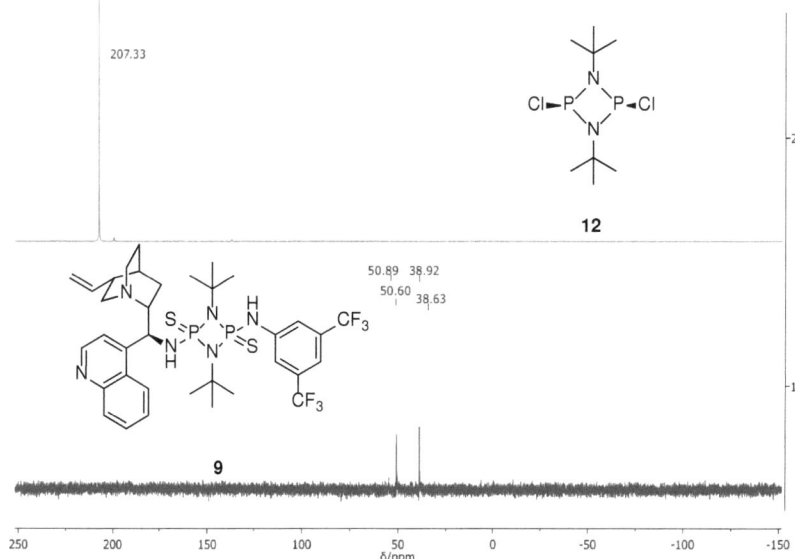

Figure 15. $^{31}P\{^{1}H\}$-NMR spectra (50 MHz) of dichlorocyclodiphosph(III)azane **12** compared to the catalyst with sulfur (s-cat) **9** in CDCl₃.

In addition, it is noticeable that the signals of the produced catalysts consist of doublets, which confirms the assumption that asymmetric catalysts are synthesized. Consequently, it can also be seen that they are oxygen-bonded or sulfur-bonded catalysts, since the signals can be seen at various chemical shifts that correspond to the literature.[13, 14]

In addition, the crystal structure of the o-cat was investigated (figure 16). First, it was attempted to recrystallize the o-cat in methanol under reflux and at room temperature, but not successful. Similarly, ethyl acetate showed no crystals. Only in acetone were crystals of the o-cats, which could also be examined. For the s-cat, no crystals have yet been formed in the aforementioned solvents. The crystal structure analysis shows an acetone complex, which gives some insight. It can be confirmed that this forms hydrogen-bonds because it forms hydrogen-bonds with a water molecule from the acetone. It can also be seen that the *cis*-configuration of the phosphazene, which is favorable for catalysis, is maintained during the course of the reaction. In addition, the tertiary amino group, which serves as a base in the bifunctional mechanism, points in the same direction as the amines on the phosphazene ring. This configuration favors the geometry for asymmetric catalysis.

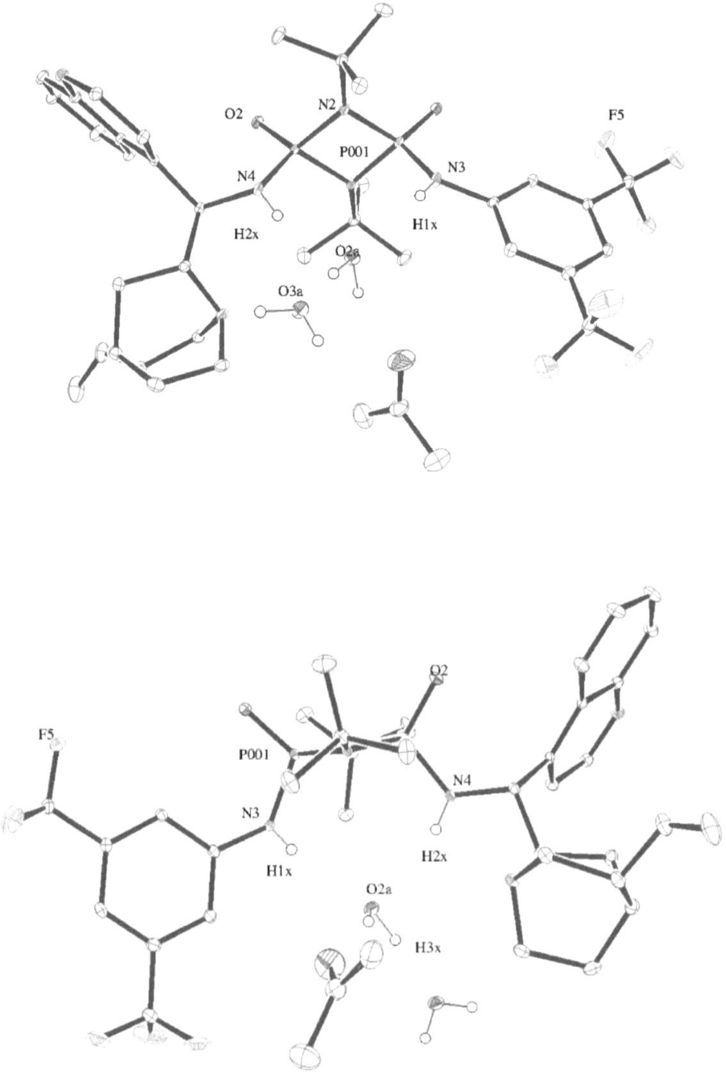

Figure 16. Crystal structures of o-cat **8** in complex with acetone. Due to the clarity the measured crystal hydrogens bound to carbons were not showed.

4.3 Asymmetric Michael Addition

As mentioned in the aim, the prepared catalysts should be tested for the asymmetric Michael addition and analysed on their catalytic activity. In this case, 2-hydroxy-1,4-naphthoquinone and β-nitrostyrene are used as substrates (figure 17).

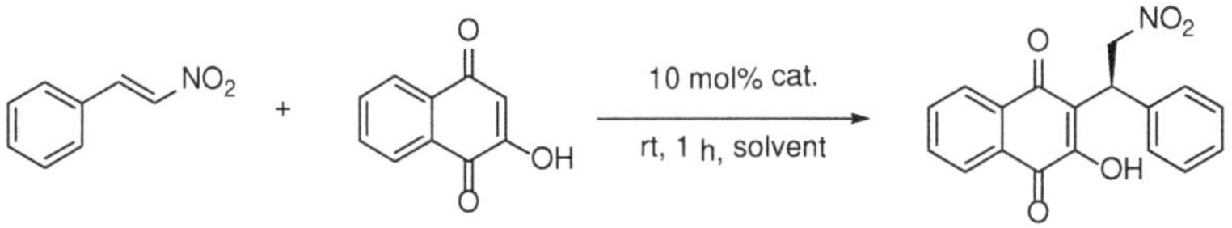

Figure 17. Tested enantioselective Michael addition of 2-hydroxy-1,4-naphthoquinone **6** to β-nitrostyrene **5** with o-cat **8** and s-cat **9** in different solvents.

All catalysis were ran for one hour at room temperature under dry conditions. The batch amount of the catalyst was always 10 mol%. The solvents have been changed. In table 1 the results are shown.

Table 1. Carried out catalysis at room temperature for one hour under inert conditions. Addition of 2-hydroxy-1,4-naphthoquinone **6** to β-nitrostyrene **5**.

Catalyst	Solvent	% Yield[a]	% ee[b] (S)
8	THF	26	36
8	**toluene**	28	**41**
8	Et$_2$O	37	32
8	DCM	77	12
9	THF	54	6
9	toluene	82	16
9	Et$_2$O	37	3
9	**DCM**	73	**27**

[a]Isolated yields. [b]Chiral HPLC Daicel OJ, n-hexane/iPrOH 50:50, 0.65 mL/min, 254 nm, rt, (S)-enantiomer: 21.04 min, (R)-enantiomer: 51.52 min.

Indeed o-cat **8** gives the best enantioselectivity in toluene, even if the yield is not the highest (41% ee, 28% yield). However, s-cat **9** achieves the highest selectivity in DCM (27% ee, 73% yield). Since no crystal structure analysis of s-cat **9** is available and thus no information about the geometry of the s-cat can be delivered, the poor enantioselectivity can be explained by a possible *trans*-configuration. For efficient and selective catalysis, both protons of the amino groups on the phosphazene ring are required for activation of the substrate (figure 18). Moreover, the enantioselectivities of both catalysts are not efficient enough, which may be explained by the bulky *tert*-butylamine groups.[5] It is possible that the hydrogen-bonding to the substrates is made more difficult.

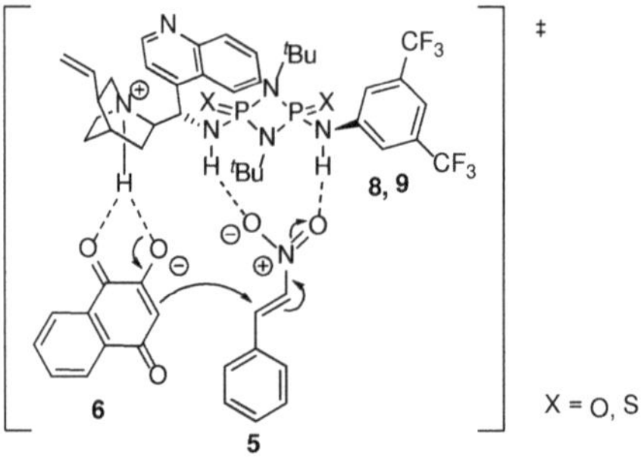

Figure 18. Postulated transition state for the enantioselective Michael addition of 2-hydroxy-1,4-naphthoquinone **6** to β-nitrostyrene **5** with o-cat **8** (X = O) and s-cat **9** (X = S).

The observations already made by *Klare* et al. that not only the chiral ligands but also the nitrogen substituents on the rings play a crucial role, in future, sterically fewer demanding substituents should be used for better enantioselectivity, without being sensitive to air and water.[5] It would therefore be possible to optimize the catalysis through changing the substitution groups on the nitrogen in the phosphazene ring.

5 Conclusion and Outlook

In this work two new chiral cyclodiphosph(V)azanes were developed. In fact, these have been successfully synthesized and used as hydrogen-bonding catalysts in the Michael addition of 2-hydroxy-1,4-naphthoquinone to β-nitrostyrene. The crystal structure of catalyst **8** meets the requirements of being in *cis*-configuration and shows hydrogen-bonding to a water molecule of the acetone. Catalyst **8** generated an enantiomeric excess of 41% in toluene and a yield of 28%. Catalyst **9** showed an enantiomeric excess of 27% in DCM, but a yield of 73% and in toluene even 82%. It is noteworthy that the enantiomeric excess is generally higher in catalyst **8**, but the yield in catalyst **9** seems to be better. A possible cause of the low enantiomeric excess for catalyst **9** could be a present *trans*-configuration. This would not be able to participate in a bidentate hydrogen bond. Consequently, the catalysts still have enough potential to be optimized in terms of enantioselectivity. Of course, reaction temperature, time, solvents, ligands, *N*-substituents could be changed.

It could be calculated how high the hydrogen-bond-donor strength is in order to compare the bonding energies. Thus, the yields could be explained by the relative strength of the hydrogen bonds.

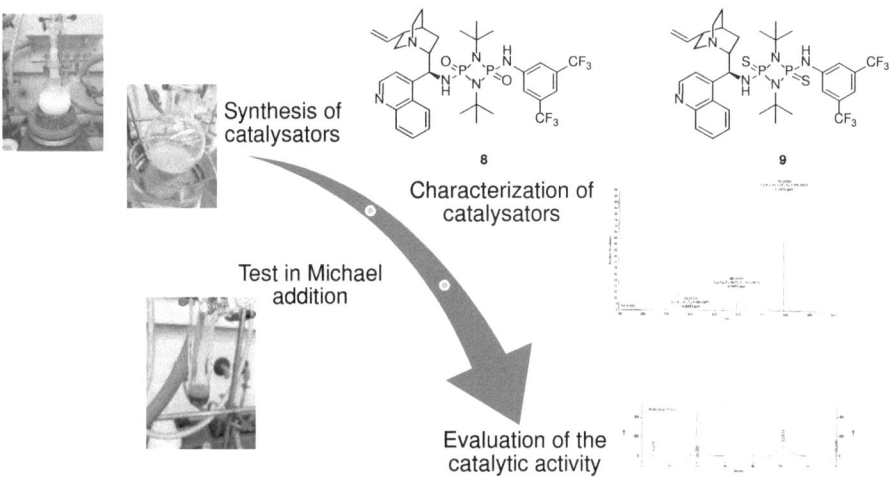

Figure 19. Schematic representation of the synthesis of o-cat **8** and s-cat **9**, their characterization, investigations and subsequent summary of the catalytic activities.

6 Experimental part

6.1 General Experimental Conditions and Analytic Methods

The chemicals and solvents used were not further purified unless stated otherwise. THF, diethyl ether, ethyl acetate and cyclohexane were separately distilled. Chemicals were purchased commercially from *Acros Organics, Fisher Chemical, Sigma Aldrich and TCI*. The protective gas used was argon from *Air Products*. THF was dried under reflux and inert gas atmosphere over sodium and benzophenone.

Working with Inert Gas

A gas/oil pump vacuum system was used for working without water or oxygen in reactions. The reaction setups were evacuated, heated and flooded with argon three times before usage.

NMR spectroscopy

The NMR experiments were recorded on a *Bruker Avance II 300 MHz* spectrometer at room temperature (^{1}H resonant frequency of 300.13 MHz). Chemical shifts δ were reported relative to tetramethyl silane (TMS). The evaluation and presentation was done with *MestReNova 6.0*. The deuterated solvent used was chloroform $CDCl_3$.

Thin-Layer Chromatography (TLC)

For TLC aluminium sheets of *Merck* (silica gel 60, F254) were used. For detection in UV light (λ = 254 nm) they were coated with silica gel and fluorescence indicator. Solvent mixtures and R_f values are listed below.

Colum Chromatography

The stationary phase was Silica Gel 60 (0.035 – 0.07 mm) from *Acros*. Flow rate was increased over pressure of a hand pump. Solvent mixtures are listed below.

High Pressure Liquid Chromatography (HPLC)

Chromatograms and spectra were recorded with a *Daicel Chiral OJ* column. *n*-hexane and *iso*-propanol were used. Both in HPLC grade. The solvent mixture was 1:1. The flow rate of 0.65 mL/min was used and the temperature was room temperature. The enantiomers of the product were determined by reference spectra.

6.2 Experimental Procedures

6.2.1 Synthesis of ligand

10 → 11

1) 1.3 eq. PPh$_3$, 1.2 eq. DIAD, 1.2 eq. DPPA
   0 - 50 °C, 14 h

2) 1.4 eq. PPh$_3$, rt, 2 h
3) H$_2$O, rt, 3 h
   THF

In a 250 ml Schlenk flask with a stir bar and a dropping funnel were added 60 ml THF under argon atmosphere. Additionally, 3.61 g cinchonidine (12.26 mmol, 1.0 eq.) and 4.22 g PPh$_3$ (16.10 mmol, 1.3 eq.) were added and stirred. The suspension was cooled down to 0 °C and at this temperature 3.0 ml diisopropyl azodicarboxylate (DIAD, 15.3 mmol, 1.2 eq.) were added. Carefully, 3.3 ml diphenylphosphoryl azide (DPPA, 15.3 mmol, 1.2 eq.) in 26 ml THF were dropped to the cloudy solution and it was stirred overnight. After heating to 50 °C for 2 h and let it cool down to room temperature, 4.58 g PPh$_3$ (17.46 mmol, 1.4 eq.) were added to the clear orange solution. It was stirred until the gas evolution was finished and 1.4 ml water were added. The solution was stirred for 3 h and then the solvent was removed under reduced pressure. The obtained oil was dissolved in 60 ml CH$_2$Cl$_2$ and 60 ml diluted HCl (10 %) were added. The aqueous phase was separated and conc. NH$_3$ was added to it until a pH value of 12 was obtained. The formed emulsion was extracted three times with 60 ml CH$_2$Cl$_2$. The organic phase was collected and dried with Na$_2$SO$_4$. The solvent was removed under reduced pressure. The crude product was purified by column chromatography with DCM/ MeOH/ conc. NH$_3$ (10: 1: 0.1). A yellow oil with a yield of 1.93 g (54 %) was obtained.

9-amino-(9-deoxy) epi-cinchonidine

habitus: yellow oil.

M (C$_{19}$H$_{23}$N$_3$):   293.41 g/mol.

yield: 1.93 g (6.59 mmol, 54 %).

^{1}H-NMR: (300 MHz, CDCl$_3$) δ [ppm] = 8.91 (d, ^3J = 4.5 Hz, 1H), 8.36 (d, ^3J = 6.7 Hz, 1H), 8.14 (d, ^3J = 8.6 Hz, 1H), 7.73 (t, ^3J = 15.4 Hz, 1H), 7.62 – 7.57 (m, 2H), 5.93 – 5.81 (m, 1H), 5.10 – 5.04 (m,

2H), 4.77 (d, ^3J = 9.8 Hz, 1H), 3.11 – 2.90 (m, 5H), 2.29 (dd, ^3J = 24.2, 8.8 Hz, 1H), 1.88 (s, 1H), 1.60 – 1.52 (m, 3H), 1.13 (dd, ^3J = 19.6, 4.5 Hz, 1H), 1.00 – 0.90 (m, 1H).

R$_f$ value: 0.38 (DCM/ MeOH/ conc. NH$_3$, 10: 1: 0.1).

Spectroscopic data agree with the literature.[10]

6.2.2 Synthesis of precursor

$$PCl_3 \xrightarrow[\text{-78 °C - 25 °C, 22 h, THF}]{\text{3.0 eq. }^tBuNH_2} \quad \mathbf{12}$$

In a 250 ml Schlenk flask with a big stir bar and a dropping funnel 100 ml THF with 8.4 ml PCl$_3$ (96.0 mmol, 1.0 eq.) were added under argon atmosphere. In the dropping funnel 50 ml THF and 30 ml $^tBuNH_2$ were submitted. At -78 °C the solution was dropped slowly within 2 hours. For another 2 hours the solution was stirred vigorously at this temperature. After stirring for 18 hours at room temperature the suspension was filtrated under argon atmosphere. The solvent was removed under reduced pressure and the solid was distilled under reduced pressure (0.1 mbar) without cooling the condenser instead it was heated with the heat gun. The temperature was set to 140 °C. The main fraction was obtained at a head temperature of 99 – 110 °C as colourless crystalline solid with a yield of 3.9 g (15 %).

1,3-Di-tert-butyl-2,4-dichloro-1,3,2,4-diazadiphosphetidine

habitus: colourless solid.

M (C$_8$H$_{18}$Cl$_2$N$_2$P$_2$): 275.09 g/mol.

yield: 3.9 g (14.2 mmol, 15 %).

^{31}P{^{1}H}-NMR: (50 MHz, CDCl$_3$) δ [ppm] = 207.33 (s).

Spectroscopic data agree with the literature.[11]

17

6.2.3 Synthesis of Asymmetric Catalyst with Oxygen

In a 250 ml Schlenk flask with a stir bar and a dropping funnel were added 1.11 g phosphazene (4.00 mmol, 1.0 eq.) and 35 ml THF under argon atmosphere. The solution was cooled down to 0 °C and within 10 min 0.62 ml $(CF_3)_2$-Aniline (4.00 mmol, 1.0 eq.) were slowly dropped and stirred for 15 min. To the colourless solution were added 0.55 ml triethylamine (4.00 mmol, 1.0 eq.) and 2 ml THF within 15 min. It changed its colour to pink and after stirring for 20 min at 0 °C the colour was slightly orange. The ice bath was removed, and it was stirred for 2 hours at rt and then overnight at 35 °C. In 20 ml THF 1.17 g Cinchonidine (4.00 mmol, 1.0 eq.) were solved and 0.55 ml triethylamine (4.00 mmol, 1.0 eq.) were added. The solution was slowly dropped within 30 min to the orange solution at -78 °C and stirred for 1.5 hours. After that, it was stirred at rt overnight. The yellow solution was filtrated under air and 100 ml THF and 0.56 ml H_2O_2 (35 %) (6.40 mmol, 1.6 eq.) were slowly dropped to the solution. For one hour it was stirred at 0 °C and at rt over the weekend. The solvent was removed under reduced pressure. The crude product was purified by column chromatography with DCM/ MeOH/ conc. NH_3 (first: 40: 1: 0.1, after second fraction: 30: 1: 0.1). A yellow solid with a yield of 0.73 g (24 %) was obtained.

cis-2-((3,5-bis(trifluoromethyl)phenyl)amino)-1,3-di-tert-butyl-4-(9-amino-(9-deoxy)-epi-cinchonidine)-1,3,2,4-diazadiphosphetidin-2,4-dioxide

habitus: yellow solid.

M ($C_{35}H_{44}F_6N_6P_2O_2$):   756.29 g/mol.

yield: 0.73 g (0.97 mmol, 24 %).

¹H-NMR: (300 MHz, CDCl₃) δ [ppm] = 8.95 (d, ³J = 4.6 Hz, 1H), 8.43 (d, ³J = 8.5 Hz, 1H), 8.16 (d, ³J = 8.4 Hz, 1H), 7.79 – 7.49 (m, 7H), 5.89 – 5.77 (m, 1H), 5.47 – 5.41 (m, 1H), 5.13 – 5.04 (m, 2H), 2.91 (d, ³J = 4.6 Hz, 5H), 2.26 (d, ³J = 7.9 Hz, 1H), 1.59 – 1.53 (m, 3H), 1.42 (s, 10H), 1.32 – 1.17 (m, 2H), 0.98 (s, 8H), 0.75 (s, 1H).

³¹P{¹H}-NMR: (50 MHz, CDCl₃) δ [ppm] = 2.22 (d), -4.67 (d).

¹⁹F{¹H}-NMR: (282 MHz, CDCl₃) δ [ppm] = -62.50.

HRMS (ESI⁺): calculated for [$C_{35}H_{44}F_6N_6P_2O_2 + H^+$]: m/z = 757.29779, found: m/z = 757.29846.

R_f value: 0.33 (DCM/ MeOH/ conc. NH₃, 30: 1: 0.1).

6.2.4 Synthesis of Asymmetric Catalyst with Sulfur

In a 250 ml Schlenk flask with a stir bar and a dropping funnel were added 1.24 g phosphazene (4.51 mmol, 1.0 eq.) and 35 ml THF under argon atmosphere. The solution was cooled down to 0 °C and within 10 min 0.70 ml (CF₃)₂-Aniline (4.51 mmol, 1.0 eq.) were slowly dropped and stirred for 15 min. To the colourless solution were added 0.62 ml triethylamine (4.51 mmol, 1.0 eq.) and 2 ml THF within 15 min. It changed its colour to pink and after stirring for 20 min at 0 °C the colour was slightly orange. The ice bath was removed, and it was stirred for 2 hours at rt and then overnight at 35 °C. In 30 ml THF 1.32 g Cinchonidine (4.51 mmol, 1.0 eq.) were solved and 0.62 ml triethylamine (4.51 mmol, 1.0 eq.) were added. The solution was slowly dropped within 30 min to the orange solution at -78 °C and stirred for 1.5 hours. After that, it was stirred at rt overnight. The yellow solution was filtrated under argon atmosphere. The solvent was removed under vacuum. 32 ml toluene and 0.32 g sulfur (9.22 mmol, 2.2 eq.) were added and the solution was stirred at

50 °C overnight. After removing the solvent under reduced pressure, the raw product was purified by column chromatography with DCM/ MeOH/ conc. NH$_3$ (first: 40: 1: 0.1, after second fraction: 30: 1: 0.1). A yellow solid with a yield of 0.22 g (6 %) was obtained.

cis-2-((3,5-bis(trifluoromethyl)phenyl)amino)-1,3-di-tert-butyl-4-(9-amino-(9-deoxy)-epi-cinchonidine)-1,3,2,4-diazadiphosphetidin-2,4-disulfide

habitus: yellow solid.

M (C$_{35}$H$_{44}$F$_6$N$_6$P$_2$S$_2$):   788.84 g/mol.

yield: 0.22 g (0.28 mmol, 6 %).

^{1}H-NMR: (300 MHz, CDCl$_3$) δ [ppm] = 8.91 (d, ^3J = 4.5 Hz, 1H), 8.45 (d, ^3J = 9.0 Hz, 1H), 8.11 (d, ^3J = 7.4 Hz, 1H), 7.76 – 7.54 (m, 6H), 5.87 – 5.76 (m, 1H), 5.53 – 5.42 (m, 1H), 5.11 – 5.03 (m, 2H), 2.97 – 2.76 (m, 5H), 2.24 – 2.22 (m, 1H), 1.59 (s, 10H), 1.40 – 1.35 (m, 4H), 1.29 – 1.17 (m, 8H), 1.10 – 1.03 (m, 3H).

^{31}P{^{1}H}-NMR: (50 MHz, CDCl$_3$) δ [ppm] = 50.75 (d), 38.78 (d).

^{19}F{^{1}H}-NMR: (282 MHz, CDCl$_3$) δ [ppm] = -62.23.

HRMS (ESI$^+$):   calculated for [C$_{35}$H$_{44}$F$_6$N$_6$P$_2$S$_2$ + H$^+$]: m/z = 789.25211, found: m/z = 789.25222.

R$_f$ value: 0.86 (DCM/ MeOH/ conc. NH$_3$, 30: 1: 0.1).

6.2.5 Asymmetric Michael addition in different solvents

In a Schlenk tube with a stir bar were added under argon atmosphere 1.2 ml solvent, 32.8 mg β-Nitrostyrene (0.22 mmol, 1.1 eq.), 34.8 mg 2-Hydroxy-1,4-naphtoquinone

(0.20 mmol, 1.0 eq.) and 0.020 mmol of the oxidized (15.1 mg) or sulfuric (15.8 mg) catalyst. The red solution was stirred for one hour at room temperature. After that, the solvent was removed under reduced pressure. The raw product was purified by column chromatography with EtOAc/ n-hexane (1: 3 in 7 g silica gel, 1: 1 in column). A red solid was obtained.

Table 2. Yields of the addition of 2-hydroxy-1,4-naphthoquinone **6** to β-nitrostyrene **5** in different solvents with two different catalysts.

Solvent	Yield with Catalyst **8**	Yield with Catalyst **9**
THF	17.0 mg (26%)	35.0 mg (54%)
toluene	17.8 mg (28%)	53.2 mg (82%)
Et$_2$O	23.9 mg (37%)	24.0 mg (37%)
DCM	49.6 mg (77%)	47.1 mg (73%)

2-Hydroxy-1,4-naphtoquinone-3-nitrostyrene

habitus: red solid.

M (C$_{18}$H$_{13}$NO$_5$):   323.30 g/mol.

^{1}H-NMR: (300 MHz, CDCl$_3$) δ [ppm] = 8.10 (dd, ^3J = 14.1, 7.6 Hz, 2H), 7.73 (dt, ^3J = 16.1, 7.5 Hz, 3H), 7.48 – 7.45 (m, 2H), 7.35 – 7.28 (m, 3H), 5.48 (dd, ^3J = 13.1, 9.0 Hz, 1H), 5.31 (dd, ^3J = 8.9, 6.8 Hz, 1H), 5.14 (dd, ^3J = 13.1, 6.7 Hz, 1H).

HPLC: Chiral HPLC Daicel OJ, n-hexane/iPrOH 50:50, 0.65 mL/min, 254 nm, rt, (S)-enantiomer: 20.35 min, (R)-enantiomer: 50.93 min.[5]

HRMS (ESI$^+$):   calculated for [C$_{18}$H$_{13}$NO$_5$ + Na$^+$]: m/z = 346.06859, found: m/z = 346.06882 (product from the reaction in DCM with catalyst **9**).

R$_f$ value: 0.29 – 0.42 (EtOAc/ n-hexane, 1: 1).

7 References

[1]   M. S. Taylor, E. N. Jacobsen, *Angew. Chem. Int. Ed.* **2006**, *45*, 1520-1543.

[2]   Y. Nishikawa, *Tetrahedron Lett.* **2018**, *59*, 216-223.

[3]   A. G. Doyle, E. N. Jacobsen, *Chem. Rev.* **2007**, *107*, 5713-5714.

[4]   A. A. Thomas, *Hydrogen Bonding Catalysis in Organic Chemistry*, **2012**.

[5]   H. Klare, J. M. Neudörfl, B. Goldfuss, *Beilstein J. Org. Chem.* **2014**, *10*, 224-236.

[6]   P. Chauhan, S. Mahajan, U. Kaya, D. Hack, D. Enders, *Adv. Synth. Catal.* **2015**, *357*, 253-281.

[7]   Y. Huang, A. K. Unni, A. N. Thadani, V. H. Rawal, *Nature* **2003**, *424*, 146.

[8]   M. S. Balakrishna, D. J. Eisler, T. Chivers, *Chem. Soc. Rev.* **2007**, *36*, 650-664.

[9]   T. Roth, H. Wadepohl, D. S. Wright, L. H. Gade, *Chem. Eur. J.* **2013**, *19*, 13823-13837.

[10]   J.-W. Wan, X.-B. Ma, R.-X. He, M. Li, *Chin. Chem. Lett.* **2014**, *25*, 557-560.

[11]   A. Bashall, E. L. Doyle, C. Tubb, S. J. Kidd, M. McPartlin, A. D. Woods, D. S. Wright, *Chem. Commun.* **2001**, *0*, 2542-2543.

[12]   N. J. Braun, Bachelorarbeit, *Entwicklung neuartiger Cyclophosph(V)azan-Wasserstoffbrückenkatalysatoren für die enantioselektive Katalyse*, Universität zu Köln, **2015**.

[13]   W. Haubold, E. Fluck, *Z. Naturforsch.* **1972**, *27*, 368-376.

[14]   http://www.chemgapedia.de/vsengine/vlu/vsc/de/ch/3/anc/nmr_spek/andere_k-erne.vlu/Page/vsc/de/ch/3/anc/nmr_spek/m_46/nmr_8_6_1/p31verschieb_m46te0403.vscml.html, 03.03.2019.

8 Appendices

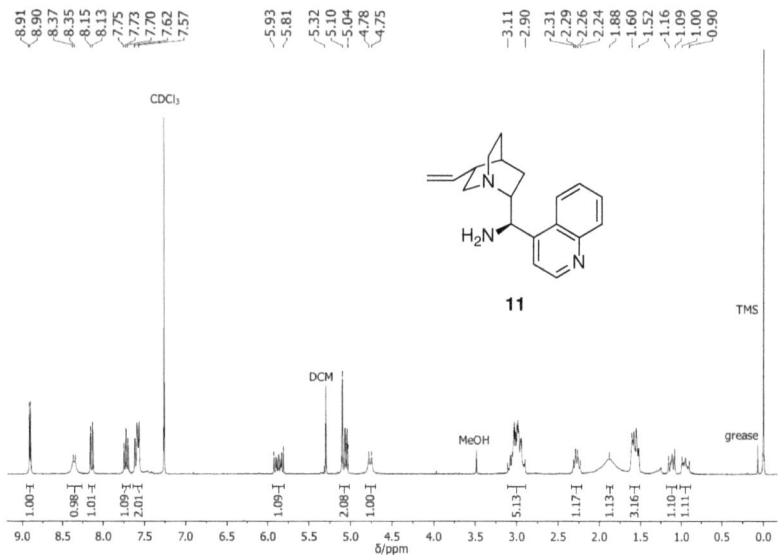

Figure 20. ^{1}H-NMR spectrum (300 MHz) of 9-amino-(9-deoxy) epi-cinchonidine **11** in CDCl$_{3}$.

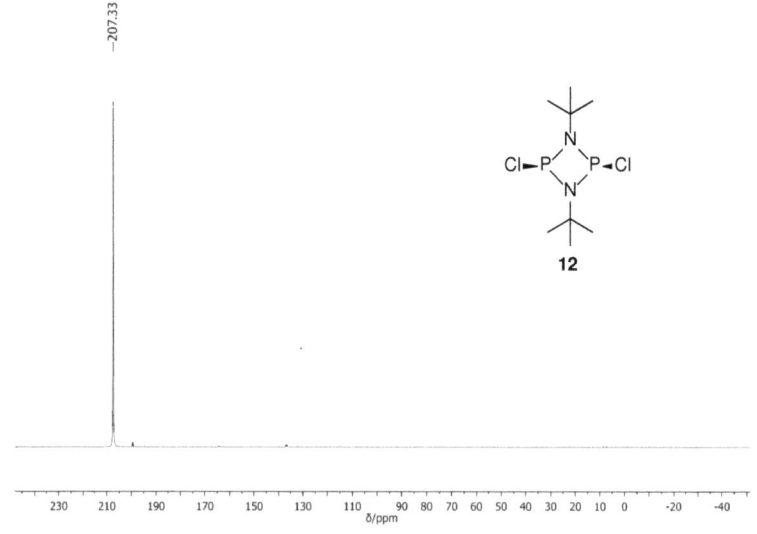

Figure 21. ^{31}P{^{1}H}-NMR spectrum (50 MHz) of 1,3-Di-tert-butyl-2,4-dichloro-1,3,2,4-diazadiphospheti-dine **12** in CDCl$_{3}$.

23

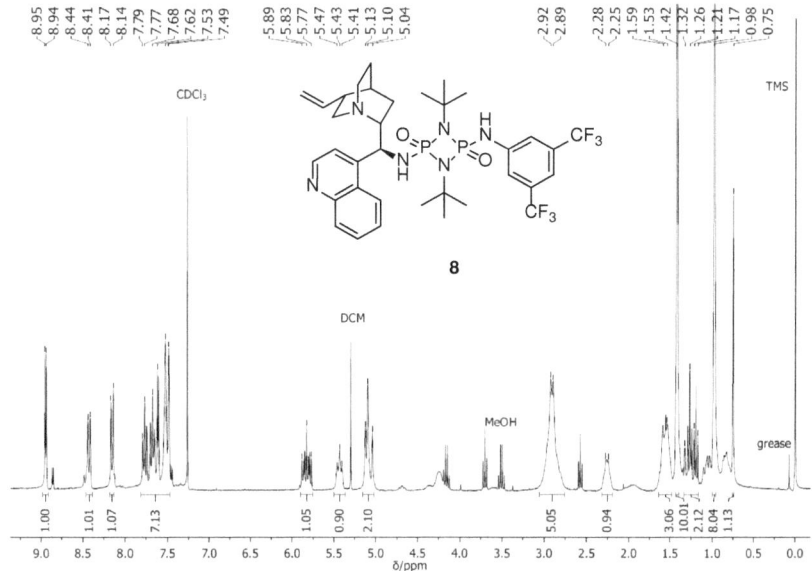

Figure 22. ^{1}H-NMR spectrum (300 MHz) of oxidized catalyst **8** in CDCl₃.

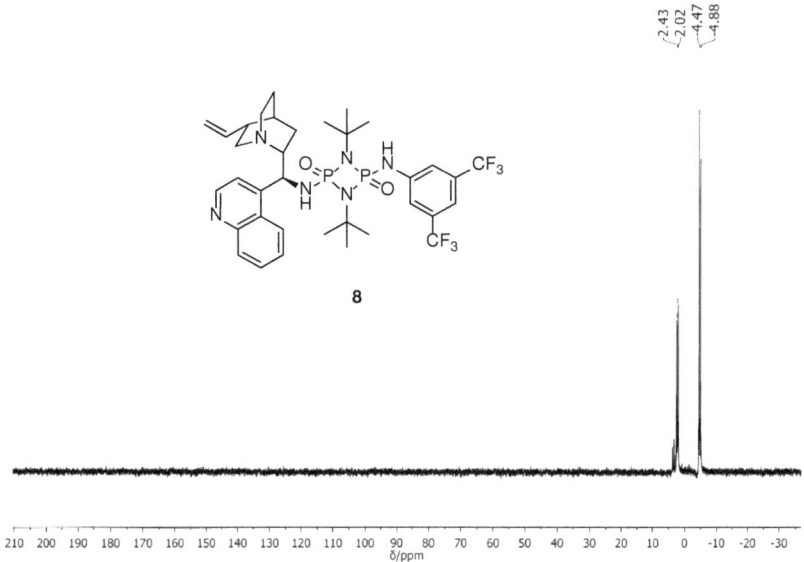

Figure 23. ^{31}P{^{1}H}-NMR spectrum (50 MHz) of oxidized catalyst **8** in CDCl₃.

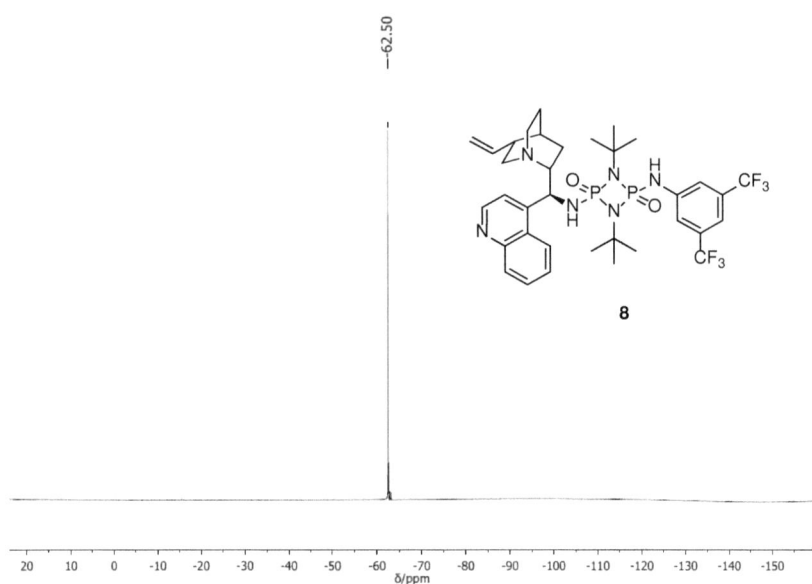

Figure 24. ¹⁹F{¹H}-NMR spectrum (282 MHz) of oxidized catalyst **8** in CDCl₃.

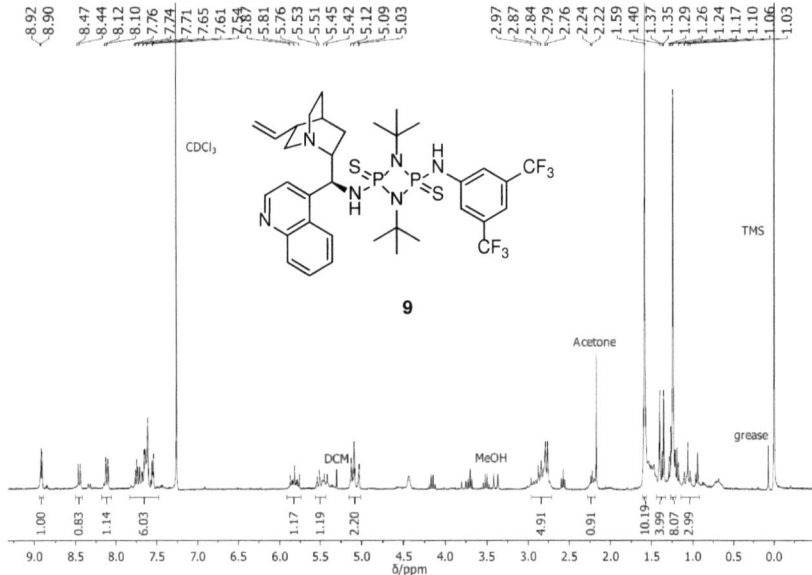

Figure 25. ¹H-NMR spectrum (300 MHz) of sulfuric catalyst **9** in CDCl₃.

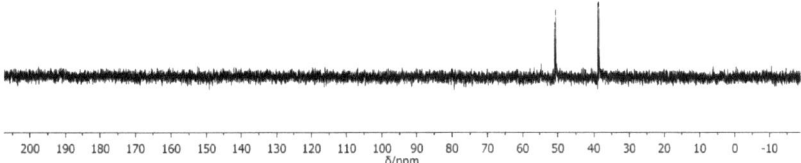

Figure 26. ³¹P{¹H}-NMR spectrum (50 MHz) of sulfuric catalyst **9** in CDCl₃.

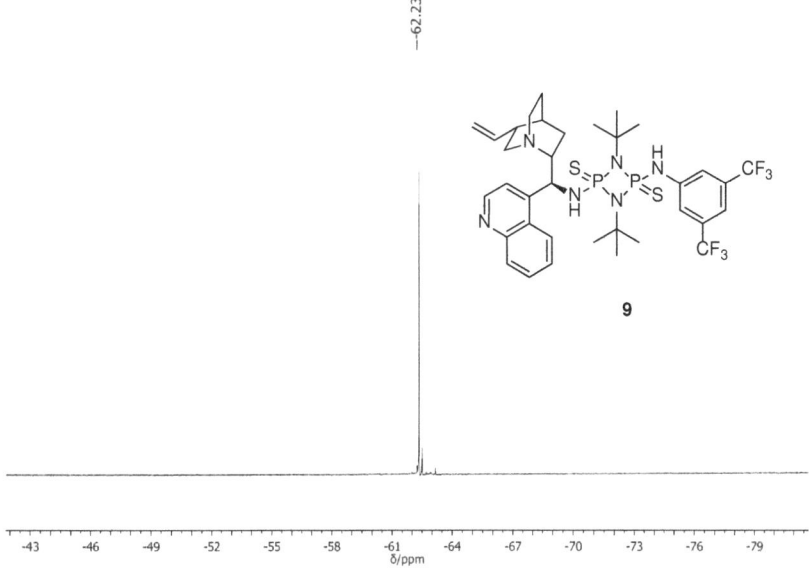

Figure 27. ¹⁹F{¹H}-NMR spectrum (282 MHz) of sulfuric catalyst **9** in CDCl₃.

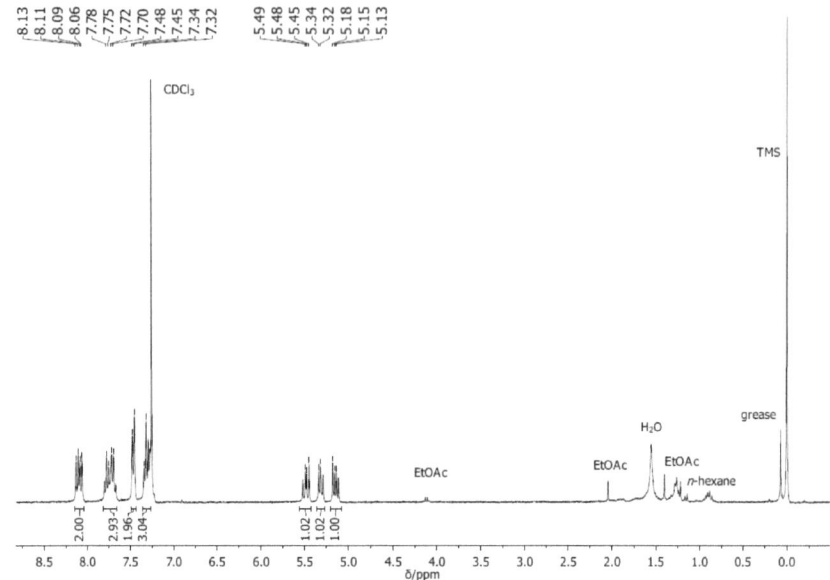

Figure 28. ¹H-NMR spectrum (300 MHz) of Michael addition product in THF with o-cat. **8** in CDCl₃.

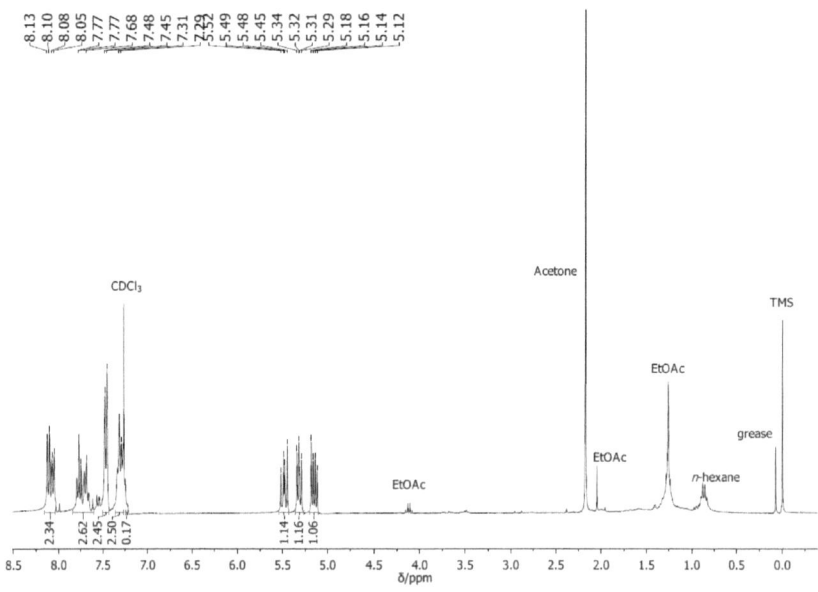

Figure 29. ¹H-NMR spectrum (300 MHz) of Michael addition product in toluene with o-cat. **8** in CDCl₃.

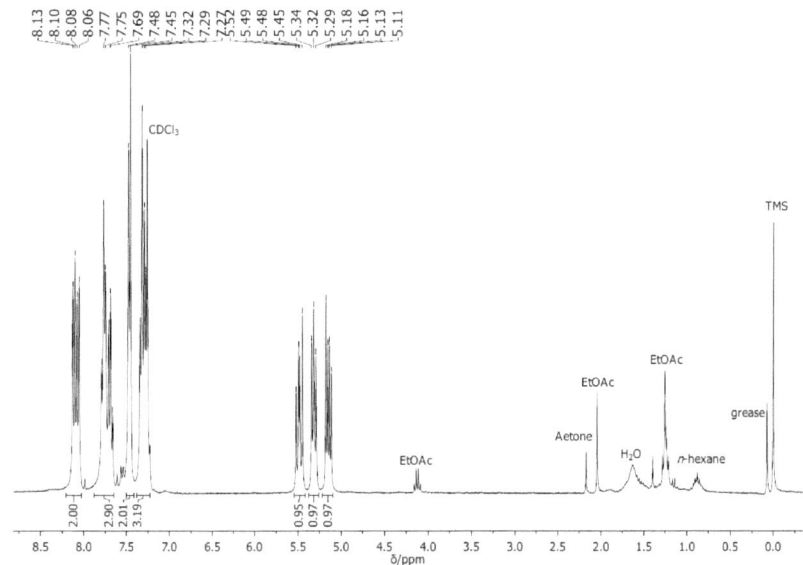

Figure 30. ¹H-NMR spectrum (300 MHz) of Michael addition product in diethyl ether with o-cat. **8** in CDCl₃.

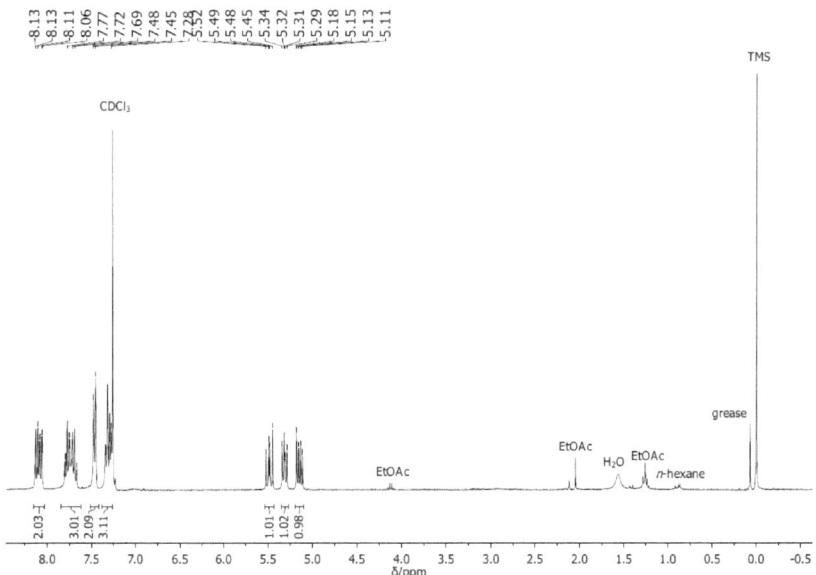

Figure 31. ¹H-NMR spectrum (300 MHz) of Michael addition product in DCM with o-cat **8** in CDCl₃.

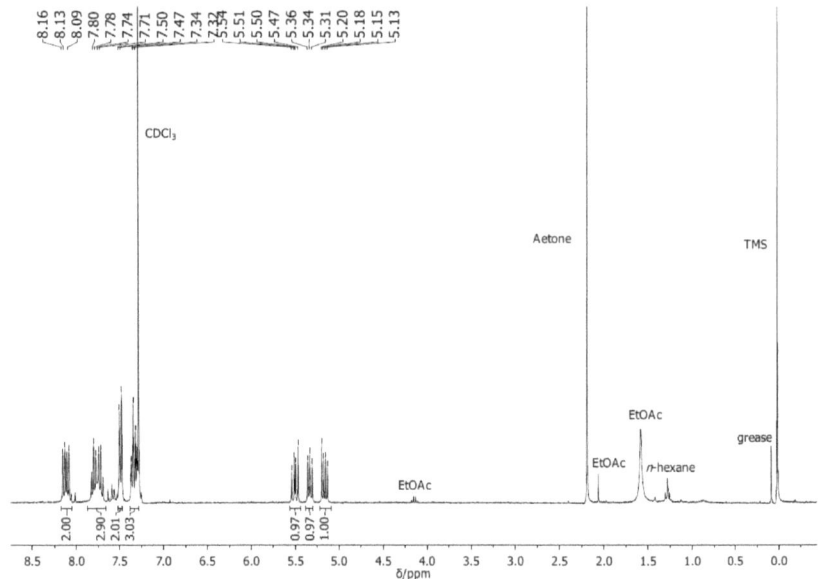

Figure 32. ¹H-NMR spectrum (300 MHz) of Michael addition product in THF with s-cat **9** in CDCl₃.

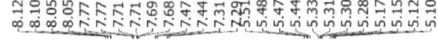

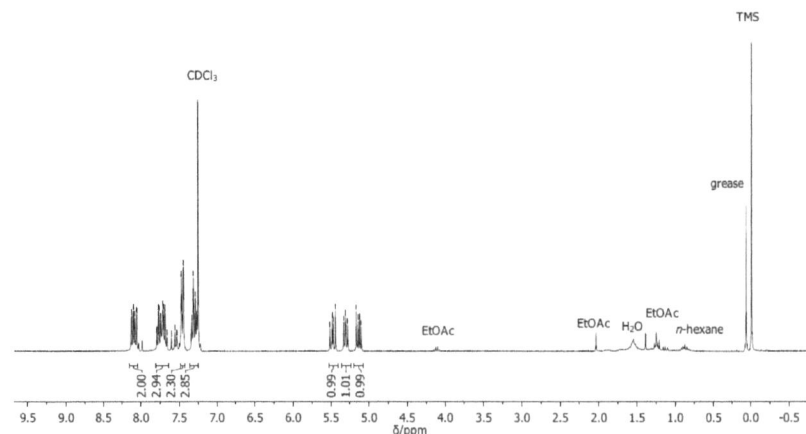

Figure 33. ¹H-NMR spectrum (300 MHz) of Michael addition product in toluene with s-cat **9** in CDCl₃.

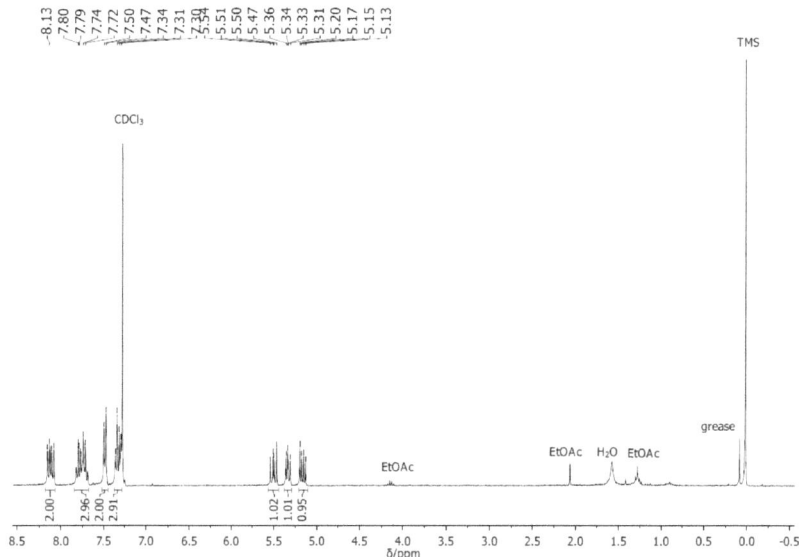

Figure 34. ¹H-NMR spectrum (300 MHz) of Michael addition product in diethyl ether with s-cat **9** in CDCl₃.

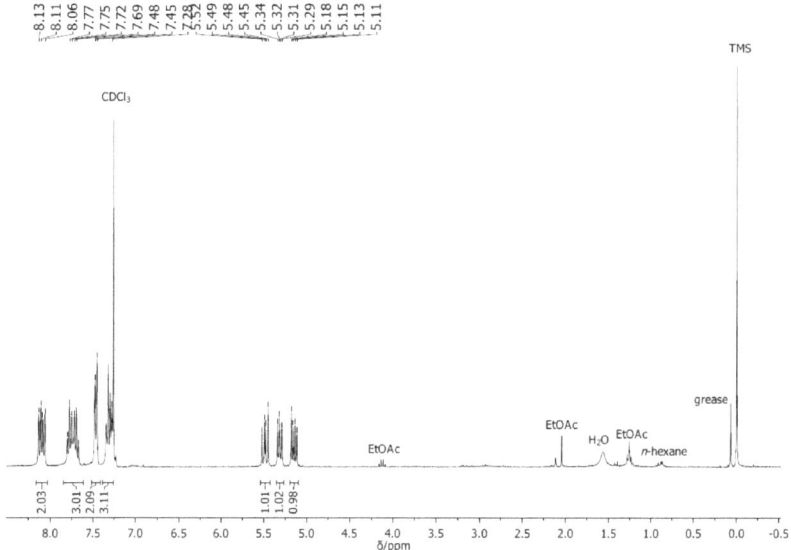

Figure 35. ¹H-NMR spectrum (300 MHz) of Michael addition product in DCM with s-cat **9** in CDCl₃.

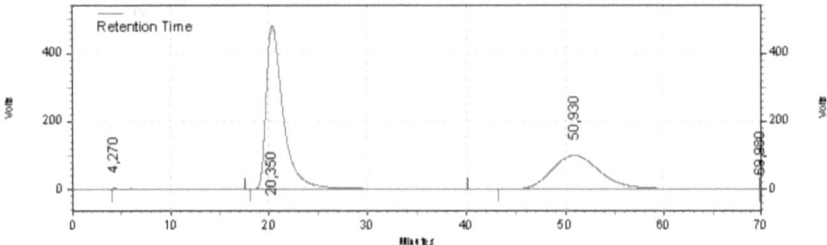

Figure 36. HPLC of the Michael product 2-Hydroxy-1,4-naphtoquinone-3-nitrostyrene **7**. Daicel Chiral OJ, *n*-Hex/*i*PrOH 50:50, 0.65 mL/min, 254 nm, rt.